# LIFE

Rebecca Woodbury, Ph.D., M.Ed.

## Gravitas Publications Inc.

# Life

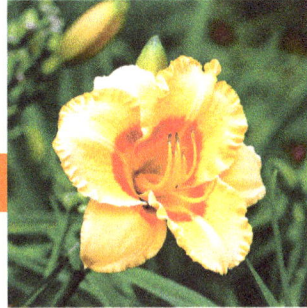

Illustrations:     Janet Moneymaker
Design/Editing:  Marjie Bassler

Life
ISBN  978-1-950415-49-6

Published by Gravitas Publications Inc.
Imprint: Real Science-4-Kids
www.gravitaspublications.com
www.realscience4kids.com

RS4K

Photo credits:  Cover and title page: Image by Michael Siebert from Pixabay; Above, Image by hartono subagio from Pixabay; p.3 Butterfly, Yuichi Kageyama on Unsplash; p.3 Leopard, Michael Siebert from Pixabay; p.3 Fish, Bruno /Germany from Pixabay; p.5 Crystal Gard on Unsplash; p.7 RÜŞTÜ BOZKUŞ from Pixabay; p.9 Nam Anh on Unsplash; p.11 Alexa from Pixabay; p.20 Lizard, shraga kopstein on Unsplash; p.20 Horse, Christine Benton on Unsplash; p.20 Koalas, Holger Detje from Pixabay; p.20 Mushroom, Benjamin Balazs on Unsplash

**Life** is everywhere.

Am I
alive too?

At the bottom of the ocean,

there is **life.**

At the top of a mountain,

there is life.

We can hike
to the top!

You do that.
I will stay here.

Living things can be **BIG**.

Living things can be small.

small

I am small.

Am I big or am I small?

Living things are different from nonliving things.

Living things are alive!

I can jump.

Living things need to eat

food to grow and move.

I need cheese!

Living things can walk,
jump, and roll.

I am glad I am not a rock!

Living things can have babies.

- Living things need food.

- Living things can grow and move.

- Living things can have babies.

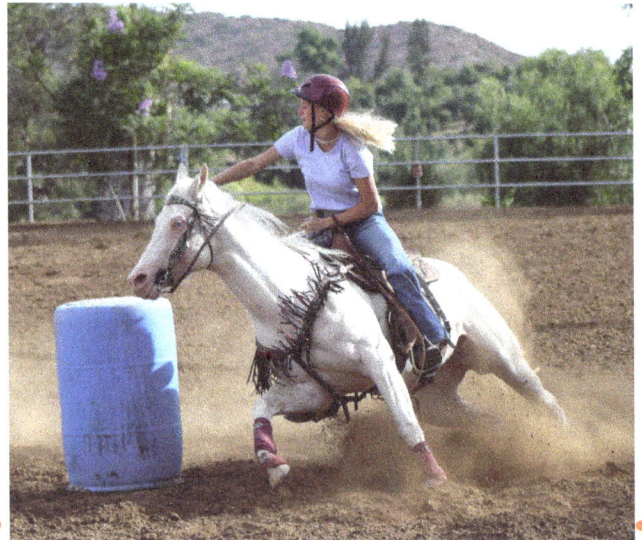

- Living things are alive!

- Living things make up everything we call life!

I am **ALIVE**!

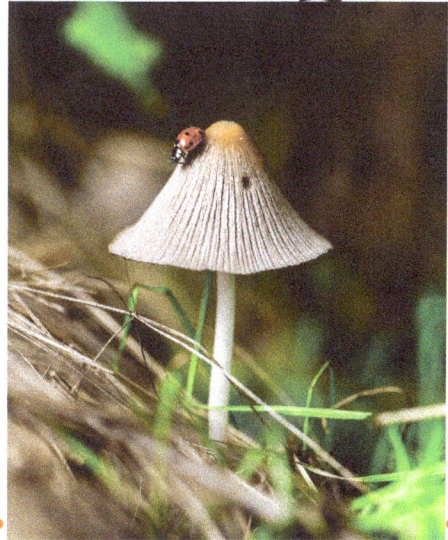

# How to say science words

**alive**   (uh-LIYV)

**life**   (LIYF)

**living things**   (LIH-ving THINGS)

**nonliving things**  (NON-lih-ving THINGS)

**science**   (SIY-uhns)

www.ingramcontent.com/pod-product-compliance
Lightning Source LLC
Chambersburg PA
CBHW040150200326
41520CB00028B/7554

9 781950 415496